Talking about:
Assisted Reproductive Technology

by Dr Trisha Prentice

MBBS BMedSci MBioeth

A small group study series from the
Centre for Applied Christian Ethics, Ridley Melbourne

Series introduction

The "Talking about: …" series was launched from the Centre for Applied Christian Ethics in 2011, and aims to provide Biblically informed guides for small group and personal study of current ethical issues. We believe that studying ethics in a small group setting can be a very challenging and valuable experience. There is a real opportunity to deeply examine important issues, and particularly to explore the ways in which faith may practically impact the way we think and live. CACE is pleased to present this series to encourage Christians to start and continue good discussions on meaningful topics.

In "Talking about: Assisted Reproductive Technology", Dr Trisha Prentice has provided a valuable contribution to this series. Issues relating to infertility and ART are increasingly common, and it is critical that Christians engage in an intelligent and faithful way with these emotive and closely felt issues.

For those intending to use this guide for small group study, or wishing to develop skills in ethical discussion more generally, CACE has also produced an introductory book, "Talking about ethics" (Acorn Press, 2011). Copies are available from acornpress.net.au

In Christ,

Justin Denholm
Centre for Applied Christian Ethics
Ridley Melbourne Mission & Ministry College

Preface

This paper has been shaped by my deep concern for friends and members of our community who are presently infertile, or have struggled with infertility. *In Vitro Fertilisation* (IVF) has become commonplace in the medical care of infertile couples. One consequence of this is that couples may not feel they have the time or space to reflect on the ethics of IVF or the implications of the process for their own life. My hope and prayer is that this paper may provide a starting point for couples (and their communities) to reflect, discuss and form their own ethical framework in the midst of what is often a time of deep grief and despair. I wish to broaden the discussion beyond the moral status of the embryo and to consider the pastoral implications of how we as a community can better love and care for one another as children of God.

Dr Trisha Prentice
April 2012

Infertility is defined as the inability to naturally conceive despite regular unprotected intercourse for at least one year. In about 40% of couples the problem originates in the man; with another 40% in the woman. Approximately 10% of couples will have infertility issues in both the man and woman, while in 10% no cause of infertility will be identified in either partner[1]. With infertility impacting an estimated 15% of Australian couples of reproductive age[2], it is likely that most people either know someone who has struggled with infertility or in fact has struggled with it personally.

Assisted Reproductive Technology (ART)[3] seeks to overcome the physical inabilities of a couple to conceive 'naturally' and produce a genetically related (to at least one parent) child. Broadly speaking, ART includes any process where eggs and or sperm are manipulated to enable conception, including *In Vitro* Fertilisation (IVF), Intrauterine Insemination (IUI), Intracytoplasmic Sperm Injection (ICSI) and Gamete Intrafallopian Transfer (GIFT). Of these, IVF is frequently considered the mainstay of ART and is used when other forms of ART have failed or are medically inappropriate. With an estimated 4 million live births around the world, and approximately 2% of births in Australia being conceived through IVF[4], it appears the technology has been largely accepted by society. The presentation of the 2010 Nobel Prize in Physiology or Medicine to Robert G. Edwards, developer of IVF, certainly appears to reflect this. It is difficult to deny the benefits of such technology. Yet underlying ethical and social concerns still remain[5].

This document aims to deal with some of the ethical considerations relevant to ART, with emphasis on IVF. For the sake of clarity, let me (the author) state my personal ethical position here at the start: I believe that IVF can be undertaken in an ethical manner that maintains a very high respect for human life. This would include (but not be limited to)

---

[1] Monash IVF. (2011). *Monash IVF Library*. Retrieved October 5th, 2011, from Monash IVF:Life starts here: http://www.monashivf.com/InformationLibrary/Why_aren_t_we_pregnant_.aspx

[2] Ibid.
This may be significantly underestimated given that much infertility remains hidden from society and some couples may not present for treatment for a number of reasons.

[3] For glossary of terms please refer to the end of the this document.

[4] IVF Babies. (2006). *IVF Babies - Information and Resources for parents of IVF babies*. Retrieved September 6th, 2011, from http://www.ivfbabies.com/

[5] For example, note the following comments by Professor Max Charlesworth at a public lecture in 1984:
> No one, save for the most die-hard Luddite would gainsay the immense human benefits that the new biotechnology has made possible, but equally no one, save the most purlind, and gun ho scientific optimist, can refuse to recognise the momentous human problems that the new technology brings in its train. Like nuclear energy, it is profoundly ambivalent with great potentialities for human impoverishment. Thus the various forms of biotechnology raise crucial moral and legal questions about the value of human life in its various stages of development, the rights of the human embryo and fetus and the new born child, the nature of human sexuality and marriage and parent-child relationships, male and female role in reproduction, the just use and distribution of medical services in a society (IVF babies for example, cost something like $50 000 each), the limits to experimentation on human life at its various stages of development, the personal identity of children born through IVF where donor gametes are used, or where a surrogate mother is involved, or where (in the future) cloning techniques are used.

As quoted in Fisher, A. (1989). *IVF The Critical Issues*. Collins Dove. P 6

limiting the number of embryos produced each cycle to the number of embryos that are to be implanted. This sort of restriction, in the midst of being on an emotional roller coaster where emphasis is placed on the outcome rather than the means, may be harder (or more impractical) than what it appears on paper. Furthermore, considerations need to be much broader than the moral status of an embryo. Once emotional, financial and relational aspects are taken into account, couples must consider whether a permissible process (ethically) is really the most beneficial (to use Paul's language of 1 Corinthians 10:23), both to the couple in question and in their relationship with God. It is my hope and prayer that outlining some of the issues may enable Christians to make more informed decisions about their own reproductive choices, or support others struggling with infertility without compromising any party's conscience.

This paper is set out in five parts. The first section introduces infertility in the Bible and God's response to it. Part 2 describes the IVF process and the role that other forms of assisted reproductive technologies may have. Part 3 addresses the Biblical and ethical concerns that the process raises. Pastoral considerations are considered in part 4. (In doing so, I can only attempt to do justice to the emotional burden and often silent suffering that infertility causes for individuals and families). The fifth section provides the foundations of a Bible study, which could be tailored to groups, individuals or couples as required.

Part 1

Infertility in the Bible

Upon completing His masterpiece of creation, God blessed Adam and Eve and gave them the mandate to "be fruitful and multiply and fill the earth" (Genesis 1:28)[6]. Yet following the Fall, creation is subjected to futility and corruption, "groaning together in the pains of childbirth" (Romans 8:18-23), while expectantly awaiting liberation and the glory that is yet to come. In this context infertility is prevalent throughout Biblical narrative. We can consider for example Abraham and Sarah (Genesis 16:1-6, 18:9-15), Isaac and Rebekah (Genesis 25:21), the anguish of Rachel (Genesis 30:1-23), and the bitter weeping of Hannah (1 Samuel 1:1-20) - to name just a few. Each example can provide us with a different insight into the nature of God and His concern for His people.

Often, these miraculous examples point to God's power and ability to achieve his purposes through His own means using fragile beings. Thus, though Sarah was past the age of childbearing, had laughed at God's promise of a son (Genesis 18:11), and had attempted to provide an heir through a surrogate (Hagar, her Egyptian servant), Abraham was to become the father of a great nation, with heirs numbering the stars (Genesis 15:5).

As such, these are exceptional accounts and cannot be used as direct promises or claims for our own lives. Though we may weep prayerfully like Hannah, God may choose not to answer our prayers in the way we desire. Sadly, some Christians falsely apply these passages to think that if they have sufficient faith and prayer they too will be blessed with children. Others may view infertility as a direct punishment for their own sinfulness[7]. Such teachings are dangerous and potentially soul-destroying.

The example of Elizabeth's righteousness makes both these errors clearer. Elizabeth and her husband Zechariah walked "blamelessly in all the statutes of the Lord" (Luke 1:6) yet until their advanced years, Elizabeth remained barren. That she miraculously conceived John the Baptist is in no way a reflection that her faith suddenly became sufficient or that a punishment had ended. Rather, God was showing His power through her. The birth of John was to fulfill the prophecy of one "to make ready for the Lord a people prepared" (Luke 1:17). Similarly, Jesus taught that there is not necessarily a direct link between sin and suffering or physical ailments. Upon seeing a blind man his diciples questioned: "Rabbi, who sinned, this man or his parents, that he was born blind?" (John 9:2-3) Jesus answered "It was not that this man sinned, or his parents, but that the works of God might be displayed in him".

---

[6] All scripture from The Holy Bible, English Standard Version™ (ESV). Copyright © 2001 by Crossway Bibles, Wheaton, IL, USA unless otherwise stated.

[7] A couple of examples of direct judgement resulting in infertility are however present in the Bible. Consider Hosea 9:14, Leviticus 20:21, Numbers 5:20-28.

We should also avoid the error of viewing biologically related children as a right in the same way that secular society appeals to human rights. Children are not something to which we can lay claim or of which we can consider ourselves worthy. Rather, the Bible is clear that they are a blessing and gift from God. This would have been especially clear to the Israelities of the Old Testament. Consider the Psalmist in Psalm 127:3-5:

> *Behold, children are a heritage from the LORD, the fruit of the womb a reward. Like arrows in the hand of a warrior are the children of one's youth. Blessed is the man who fills his quiver with them! He shall not be put to shame when he speaks with his enemies in the gate.*

There is a sense in which the children belong to God, rather than a possession rightfully claimed. The 'reward' is better understood as wages[8]. That is, for the Israelites the blessing of children provided an inheritance and wage in a time where there was no form of social security. Furthermore male children offer protection of the family unit or city[9]. The Israelites were also living in light of the Abrahamic promises. For the Israelites there was a much stronger connection to the land in which they lived and an identity as a distinct God-fearing nation. The fulfillment of God's covenant with Abraham and the growth of His nation was reliant on fertility and procreation for its success or fulfillment. Thus in Deuteronomy 7:12-15 the Israelites are told:

> *And because you listen to these rules and keep and do them, the LORD your God will keep with you the covenant and the steadfast love that he swore to your fathers. He will love you, bless you and multiply you. He will also bless the fruit of your womb and the fruit of your ground, your grain and your wine and your oil, the increase of your herds and the young of your flocks, in the land that he swore to your fathers to give you. You shall be blessed above all peoples. There shall not be male and female barren among you or among your livestock. And the LORD will take away from you all sickness, and none of the evil diseases of Egypt, which you knew, will he inflict on you, but he will lay them on all who hate you.*

Note that God's literal blessings of land and fertility for those who follow in his ways are made on a national level rather than as promises to individuals. As we have seen above, sickness and infertility still existed, and exist today, irrespective of an individual's faith or sinfulness. As Christopher Wright notes: "the connection between faith, obedience and material blessing is neither instant and automatic, nor universally experienced"[10]. While God continues to delight in blessing those who are faithful to Him (though again not necessarily instant, automatic or universal) Christians no longer relate to the land or express nationhood in the same way as the Israelites of the Old Testament. Through faith in Jesus Christ, the mediator of a new covenant (Hebrews 9:14), we are now spiritual offspring of Abraham, "heirs according to promise" (Galatians 3:29). The emphasis has changed from an expectation to be "fruitful and multiply" in a physical sense to growing His kingdom through spiritual heirs. The apostle Paul understood this. Though single and

---

[8] Goldingay, J. (2008). *Psalms 90-150* (Vol. 3). Baker Academic. Pg 503-504
[9] Ibid.
[10] Wright, C. (1996). *Deuteronomy*. (R. J. Hubbard, Ed.) Peabody, Massachusetts: Hendrickson. Pg 117-119.

without children he stated: "[f]or though you have countless guides in Christ, you do not have many fathers. For I became your father in Christ Jesus through the gospel. I urge you, then, be imitators of me"(1 Corinthians 4:15-16). Paul appreciated the blessings that come from being a father to many spiritual children[11]. This is not to say that biological children are any less of a blessing or gift today, but they are no longer the fulfillment of God's promises in the same sense that they were for the Israelites. We need to accept that God may and does bless us in different ways and at different times.

Suffering

Understandably, infertility results in a great sense of loss and subsequent suffering. With mankind's rebellion against God (Genesis 3) pain, suffering and death enter the world. Subsequent generations could not and will not escape the consequences of the Fall – irrespective of individual sinfulness or righteousness[12] – until the final restoration of God's people and the creation of a new heaven and new earth (Revelation 21:1). Only then will "He will wipe away every tear from their eyes, and death shall be no more, neither shall there be mourning, nor crying, nor pain anymore, for the former things have passed away" (Revelation 21:4).

While suffering should be expected and is inevitable, it is not always futile. Rather, in order to share in Christ's glory, we must first share with Christ in his sufferings (Romans 8:17). Through suffering and hardship may come the development of character and perseverence (Romans 5:3-4, James 1:3). It can refocus our sights on God and remind us of our dependence on him. It is also limited to this lifetime (Revelation 21:4).

That we are growing in character and perseverance may be of little comfort during times of suffering and trials. Instead there is often the desire to understand why God allows such suffering to occur. The book of Job provides some insight. Though God-fearing, blameless and upright (Job 1:1), God allows Satan to cause extreme suffering to Job to which none of our sufferings can compare. While his friends unfairly rebuke him and even his family advise him to curse God, Job holds fast in trusting God – but understandably asks why. God's response (Job 38-41) is to rebuke Job, reminding him of His vast power and ways that are beyond all understanding and intelligence. Like Job, we must stand before God with humility, trusting his unfailing love and sovereignty even though many things including suffering are beyond our full understanding.

Not only is our suffering not in vain, but we do not endure it alone. God is the 'God of all comfort' (2 Corinthians 1:3) who can empathise with our temptations and suffering (Hebrews 2:18, 4:15). For on our behalf, the sinless son of God would bear our sins on the cross and experience the ultimate suffering in order that our relationship with God could be restored.

---

[11] See also: Piper, J. (2007, April). *Single in Christ: a Name Better Than Sons and Daughters*. Retrieved November 2011, from Desiring God: http://www.desiringgod.org/resource-library/sermons/single-in-christ-a-name-better-than-sons-and-daughters#/watch/full

[12] Again I must stress the unhelpfulness of correlating particular sufferings to direct sinfullness.

Seeking Restoration

For some, IVF may lessen the suffering of infertility through the creation and celebration of life. In this light, treating infertility can be seen as a restorative process – that is, something that seeks to return the body to its (God's) intended role and functionality. If we readily acknowledge that children are a gift and a blessing from God (Psalm 127:3), and view healing as a loving and compassionate virtue, then why is the use of IVF in Christian circles considered a 'grey zone' with divided opinions? We must first examine the process and question whether the means justify the ends.

Part 2

What IVF involves

The IVF process is individually tailored to each woman depending on her underlying reason for infertility. The following steps are only a guide to what the process is likely to involve.

1. Pituitary suppression

From about day 21 of the menstrual cycle a hormone medication (such as a Gonadotrophin releasing Hormone [GnRH] agonist) is provided to cease normal release of the pituitary hormones Leutinising Hormone (LH) and Follicle Stimulating Hormone (FSH)[13]. These hormones are involved in the normal maturation and ovulation of an egg from the ovary. Thus this step enables the doctors to control the ovulation cycle.

2. Ovarian stimulation

At the beginning of the patient's menstruation an ultrasound and blood tests are performed to assess suitability for stimulation. Depending on a patient's individual protocol, fertility drugs are then provided for about a week (up to two weeks). These fertility drugs stimulate the growth of multiple follicles rather than the usual single follicle produced in a normal cycle. Ultrasounds and blood tests monitor the process. Approximately 5-10% of women do not produce sufficient follicles through this process to proceed with IVF[14]. Other women will have to cease treatment at this point due to the risks of Ovarian Hyperstimulation Syndrome (OHSS). OHSS is a potentially life-threatening syndrome triggered by ovulation that may result in the build up of fluid in the abdomen and chest. Overall cancellation of treatment occurs in up to 20% and is more likely to occur in those older than 35 years.[15]

3. Egg retrieval

Once sufficient follicles are present, hormonal stimulation (GnRH agonist and FSH injections) is stopped. An injection of another hormone – human chorionic gonadotrophin (hCG) – is given, which results in final maturation of the eggs. 36 hours later the follicles will be retrieved through ultrasound-guided needle aspiration through the back wall of the vagina up to the ovaries. On average 8-15 eggs are retrieved[16]

---

[13] All medications used in the IVF process can have significant side effects. These can significantly add to the burden and stress often experienced during IVF processes.
[14] City Fertility Clinic. (2009). Retrieved September 15th, 2011, from IVF Process - IVF Treatment: http://www.cityfertility.com.au/IVF-Process
[15] Gurevich, R. (2011). Retrieved September 5th, 2011, from About.com Fertility: http://infertility.about.com/od/infertilitytreatments/ss/ivf_treatment_5.htm
[16] Ibid. http://infertility.about.com/od/infertilitytreatments/ss/ivf_treatment_6.htm

The partner's semen sample will be collected on this day (if suitable). Donor sperm may also be used.

4. Fertilisation

The sperm are washed and concentrated, then added to the eggs. In certain cases of male infertility such as poor motility, a process called Intracytoplasmic Sperm Injection (ICSI) may be used to inject a single sperm directly into the egg.

Approximately 60-70% of the eggs will fertilise (becoming a zygote – the earliest developmental stage of an embryo) assuming normal sperm quality. A fertilised egg or zygote does not always result in a viable embryo.

Some couples who are carriers of genetic disorders may consider preimplantation genetic diagnosis of embryos prior to embryo transfer. In this case, one or two cells from the early embryo (called a blastocyst at this stage – day 4-6 post fertilisation) are removed and examined for specific disorders. Genetically affected embryos are then discarded.

5. Embryo transfer

On day two or three (or later if preimplantation genetic diagnosis is utilised) post egg retrieval, one to two embryos will be transferred into the uterus using a special catheter. Generally the embryos deemed the 'best quality' will be inserted.

A form of progesterone is then provided daily to optimise the uterine lining (endometrium) for implantation.

Though additional embryos may have been produced, these will not be inserted, in order to avoid the increased risks associated with multiple births. Couples will be given the option of freezing remaining suitable embryos (cryopreservation). 'Suitable' embryos do not show any evidence of fragmentation (cell breakdown) or abnormal development. Couples may subsequently choose to thaw embryos for future cycles should this cycle not be successful or if further children are desired. Inevitably some embryos will not survive this process. If further cycles or children are not desired, the couple may choose to donate surplus embryos[17]. Other embryos may be used for research. Yet other embryos will be allowed to perish.

Some couples may choose to fertilise and produce only sufficient embryos required for one cycle at a time (i.e. up to two embryos) to avoid the ethical dilemmas surrounding surplus embryos (which we will subsequently explore). Statistically this is likely to

---

[17]This 'adoptive' process raises further questions and considerations including identity. A retrospective study performed by Monash University's IVF clinic using data spanning over 10 years revealed that almost 90% of parents were likely to discard surplus embryos rather than donate them, being unwilling to "consider full siblings of their existing children living with other families, or the risk of unwitting sibling intermarriage". See: Kovacs, G. T. (2003). Embryo donation at an Australian university in-vitro fertilisation clinic: issues and outcomes. Retrieved January 27, 2012, from The Medical Journal of Australia: http://www.mja.com.au/public/issues/178_03_030203/kov10329_fm.html
A full analysis of the adoption process cannot be provided in this document.

expose the patient to increased cycle numbers with the associated risks of anaesthetic, potential Ovarian Hyperstimation Syndrome and the emotional stress (both individually and as a couple) associated with the infertility process. Couples unwilling to allow embryos to perish may also have to face the difficult decision of whether to implant an embryo with genetic abnormalities. While some of these embryos will perish during the implantation process or subsequently miscarry, some will continue to develop to term and may or may not demonstrate any phenotypic abnormalities. The long term significance of some of these genetic aberrations is unknown (it is quite possible that many of us carry small genetic 'abnormalities' of which we are completely unaware).

Currently, freezing unfertilised eggs (rather than embryos) is not routinely performed as the eggs are often structurally damaged during the cryopreservation process. This reduces the success rate of future IVF. However technological advancements may soon mean that this will be an option, potentially eliminating one of the 'grey areas' in IVF.

6. Luteal phase

This is the two-week period between embryo transfer and testing for a successful pregnancy via a blood test. It is understandably an anxious time for couples trying to conceive. Should a successful conception take place, the pregnancy will be monitored closely for miscarriage and ectopic preganancies, the former occuring in up to 35% of women over 40[18].

---

[18] Ibid. http://infertility.about.com/od/infertilitytreatments/ss/ivf_treatment_10.htm

Part 3

The issues

With this background, let us now explore some of the ethical questions and concerns surrounding IVF and other assisted reproductive technologies.

Personhood

Scientifically, the penetration of the egg by the sperm results in a complete set of genetic material, which, unless the process is halted by one thing or another, will continue to form a new human being. The same cannot be said about a sperm or egg, even if lying in the same petri dish, until they fuse together. Much of the ethical debate surrounding IVF stems from questioning whether this first human cell (called a zygote), with a complete DNA 'blueprint', is worthy of our concern and protection.

Some, including many prominent Christian ethicists[19] note that this early embryo[20] still has the potential to naturally or artificially be split into identical twins until approximately 14 days, when the "primitive streak" develops. This embryological marker forms the beginnings of the spinal column and central nervous system. How, they would argue, can we consider the embryo an individual when in fact it may develop into two distinct individuals? However, the possibility of twinning does not undermine the organisational integrity of the pre-twinning embryo. This situation could be likened to cloning. Though the orginal being and the clone share the exact same DNA (putting aside the natural genetic mutations and environmental influences that occur) it is difficult not to consider the two beings as distinct individuals[21].

This two-week mark also corresponds with approximately the time of implantation. The embryo consists of approximately 64 cells at this stage and is technically known as a blastocyst. Implantation requires the outer cells of the blastoyst to invade the uterine wall where it will receive the nourishment required for subsequent growth and development. Some thus feel that until implantation takes place, the embryo cannot be considered a potential human being as there is potential for the process to be halted[22].

These, however, remain arbitary time frames. That the development of a human being may be halted at any point from the time of conception does not remove its potential prior to that point. That is, we would be hard pressed not to consider a 36 week foetus an individual prior to the unfortunate events of being born dead from unknown causes. We

---

[19] Including Dr. Denise Cooper-Clark, Alan Nichols and previous Primate Archbishop Peter Carnley.
[20] Some literature will refer to the zygote up until 14 days as a pre-embryo. However, I will use the terminology of embryo for simplicity.
[21] IVF The Critical issues, Anthony Fisher, Melbourne; Collins Dove 1989. P 138
[22] That the embryo has no moral value until 14 days after conception was accepted by the Warnock Committee – a Brittish ethics committee led by Mary Warnock in 1984 to inquire into Human Fertilisation and embryology. Their work has provided the ethical foundations for many assisted reproduction organisations.

must question whether moral reasoning has been applied or whether we have created definitions to suit our perceived needs or desires.

Of note, within bioethical debate, is the distinction often made between an individual human being and a person. The former refers to the biology, that is to say, that we are of the species *homo sapiens*. In contrast, personhood is considered of more moral significance. The qualities or characteristics that define personhood are hotly debated but may include (as a minimum) qualities such as self-awareness, the ability to reason and an ability to relate to others. One problematic element of personhood, according to definitions based on capabilities, is that it implies our moral worth may vary with the trajectory of time. That is, at the very earliest point of life and as we near death we may not be able to relate to others in a meaningful sense. Yet in the prime of our life our capabilities to contribute to society will be much greater. Are we therefore to say that an intelligent being in the prime of his/her life has more moral value than a person approaching death, or a newborn baby?

The Bible does not make such a distinction between being a human being or a person. Nor does it specify a point in time when we develop moral significance. Frequently referred to is Psalm 139.13-16:

> *For you formed my inward parts;*
> *you knitted me together in my mother's womb.*
> *I praise you, for I am fearfully and wonderfully made.*
> *Wonderful are your works;*
> *my soul knows it very well.*
> *My frame was not hidden from you,*
> *when I was being made in secret,*
> *intricately woven in the depths of the earth.*
> *Your eyes saw my unformed substance;*
> *in your book were written, every one of them,*
> *the days that were formed for me,*
> *when as yet there was none of them.*

We must keep in mind that this is not a scientific piece of writing. Rather it is a poetic piece that reflects on the fact that God knows us and cares about us before we are born. While we can't use it to pinpoint when our personhood begins, we can deduce that at some point before birth God knows us . The story of John the Baptist within the womb reiterates this. In Luke 1:13-17 the angel Gabriel proclaims the life that John will follow, before he is even born. Furthermore we are told that John will be "filled with the Holy Spirit, even from his mother's womb" (Luke 1:15).

That we can be known by God, and that God is mindful of us (Psalm 8) is no small theological truth. It implies that we have value in God's eyes prior to any self awareness or self-consciousness - before we demonstrate our ability to contribute to society.

Furthermore, our dignity and value from a Christian perspective are derived from being created in God's image (Genesis 1:26-27). What this means in detail has been a regular topic of theological debate. What we can say is that from Genesis this creative action of God distinguishes us from other animals, as demonstrated by our God-given dominion to rule over them in a way that reflects God's dominion. It reflects God's concern for his creation and defines our unique ability to relate to God. It grounds us as beings with the capacity to make moral choices and distinguish the difference between right and wrong. To be made in God's image is an intrinsic quality present from the point of conception and is thus irrespective of cognitive capacity, apparent worth to society, or the subjective notion of quality of life'.

Sanctity of life

If embryos can be considered persons with God-given intrinsic value then it must be asked whether causing their demise can ever be justified. This is often referred to as the Sanctity of Life principle. God certainly places a high value on human life (Deuteronomy 20:19 and Numbers 35:31, for example). Biblical exceptions appear largely limited to holy war or as dictated by God in specific circumstances to maintain the purity of Israel (e.g. Deuteronomy 24:7).

However the Bible calls us not just to a high respect for human life but also specifies our obligation to defend and protect the 'defenceless' or most vulnerable in society. Deuteronomy 10:17-18 demonstrates the justice of the 'God of gods' and 'Lord of lords' who defends and shows mercy to the fatherless, widow and the aliens amongst the Israelites[23]. His people are called to show the same love, compassion and mercy to the most vulnerable of society or – alternatively - face His wrath (Deuteronomy 22:21-27, Zechariah 7:81-14). This call for compassion is continued in the New Testament (Luke 14:13, James 2:2-6) and is equated with righteousness (James 1:27, Isaiah 1:27).

If, at a minimum, we can agree that an embryo is a potential person, then it follows we can view embryos as some of the most vulnerable in our society and worthy of protection. Before utilising IVF, couples must therefore satisfy themselves with answers to the following questions: Is it ethical to subject embryos to cryopreservation knowing that some will not survive the process? Is human life 'bankable'? Is it ever right to justify the means by the ends? Arguably, where there is doubt we should err on the side of caution.

Conscience

Given that the Bible doesn't directly answer all our questions surrounding asisted reproductive technologies, we would be wise to heed Paul's advice regarding the conciences of others. In Romans 14 Paul describes weak Christians – likely Jewish Christians – who, though firmly believing, are sensitive with respect to their freedom to maintain or reject particular rituals or behaviours that are by no means central to their

---

[23] Wyatt, J. (2004). *Matters of Life and Death.* Leicester, England: Inter-Varsity Press. Pg 68-69.

salvation[24]. The strong (who do not share these sensitivities), in response, "have an obligation to bear with the failings of the weak", not pleasing themselves but rather "pleas[ing] his neighbour for his good, to build him up" (Romans 15:1-2). The end goal is that the "God of endurance and encouragement grant you to live in such harmony with one another, in accord with Christ Jesus, that together you may with one voice glorify the God and Father of our Lord Jesus Christ" (Rom 15:5-6).

This should not limit our engagement in discussing the morality of procedures such as IVF. Rather we should ensure our engagement is to build one another up and not to pass judgement. Furthermore, those with weak consciences - that is, those who are ethically concerned – must take care when proceeding with IVF for "whatever does not proceed from faith is sin" (Romans 14:23). One hears of couples who, having proceeded with IVF and rejoicing in the new life before them also find themselves doubting their own ethical integrity during the process. Some may even admit that they would not do it again. This raises another question: does the end – the creation of life – ever justify the means?

Means vs ends

A prominent Christian IVF physician recently made the heartfelt comments: "I have slowly found it untenable to put my intellectual concepts of the beginning of life over the needs of the sick. Living with unease with some aspects of IVF is the price I pay for being able to work healing in the pain of infertility"[25]. That is, even though he did not feel fully comfortable with IVF technology's approach to the beginning of life, in light of the existing ambiguities, he felt that his call to heal, restore and enable God's gift of life was a greater mandate. Certainly one cannot deny his noble intentions. Yet he has (in my view) compromised his conscience for an outcome. His rationale implies that the ends justify the (ethically ambiguous) means. This form of justification is commonly seen in a secular utilitarianism that seeks the greatest good for the greatest number of people irrespective of the means used to achieve it. In contrast, from a Christian perspective, we acknowlege that actions can be right and wrong in themselves and should be judged accordingly. Paul picks up on this when he writes "What shall we say then? Are we to continue in sin that grace may abound? By no means! How can we who died to sin still live in it?" (Romans 6:1-2).

Does biology have moral significance?

All forms of artifical/assisted reproduction[26] fertilisation occur without the need for sexual intercourse. Some argue[27] that this is outside God's creation mandate[28]. That is, according to this view, babies are 'made' in a laboratory rather than 'begotten' through

---

[24] Stott, J. (1994). *The Message of Romans.* Leicester, England: Inter-Varsity Press. Pg 355-375.
[25] Chenoweth, J. (2011). What is it to be human?: The view from IVF practice - a personal reflection. (D. J. Foley, Ed.) *Luke's Journal , 16* (2). Pg 12-13
[26] Including IVF, Artificial insemination – donor insemination (DI) and that utilising the husbands sperm (AIH) – Intra-Cytoplasmic Sperm Injection (ICSI), Gamete Intra-Fallopian Transfer (GIFT), Zygote Intra-Fallopian Transfer (ZIFT), Surrogacy.
[27] Commonly Roman Catholics.
[28] Those who hold this view may even consider the act of masturbation to obtain the sperm sample as problematic.

the loving union of parents being 'one flesh' before God. The term begotten reflects the idea formulated by the early church fathers, dating back to the Nicene Creed. That the Son was begotten reflects that He was 'of one substance with the Father'. What we beget we cannot control but rather receive as a gift from God. In contrast, what we make is unlike us – not sharing in our human nature – and is a product of our will that is at our disposal[29].

Certainly, children born through ART share in our humanness. Moreover, with such variable success rates of ART children conceived through these processes are very much still considered a gift or blessing by most parents. However we must be aware that ART can encourage us to think of children as a product of our will or even a possession rather than a gift. The ability to select an embryo for implantation based on its genetic makeup further enhances this issue.

As stated previously, IVF can be considered a *restorative process* if occuring within the covenant of marriage and when utilising the couple's own egg and sperm to overcome infertility. In contrast, gamete donation (i.e. egg or sperm donation) and surrogacy are more problematic for many Christians; these processes step beyond restoration, altering God's creative intent by changing the relationship between a parent and child through the introduction of a third party[30]. Some have even ventured so far as to call the involvement of a third party a form of adultery. This is too strong a term as there is no human relationship outside of the marital relationship[31]. However the introduction of genetic material from a donor still has the potential to place stress on the marital unit and potentially cause identity concerns for the resulting child. That many parents are frequently hesitant to inform their children (or others) of their conception through donated sperm may reflect perceived differences in the relationship both within the family and within society. Parents choosing to utilise sperm donation should maintain openness and honesty with their children. They should also consider in advance how and when the child will be informed about his/her genetic lineage, and how they – the parents – will respond should the child wish to trace the donor's details.

Donated gametes raise other issues. We are fortunate in Australia that sperm donation is a highly regulated process. In Victoria sperm donors must be willing to release identifying information to the conceived child and are limited in the number of donations they can make. They do not, however, have any legal rights or responsibilities in relation to the genetically related child[32]. This is in contrast to countries such as America where donors remain anonymous and where at least one case of a sperm donor parenting 150 children has been identified[33]. Questions have been raised regarding how we weigh up the rights

---

[29] Wyatt, J. (2004). *Matters of Life and Death*. Leicester, England: Inter-Varsity Press. Pg 89-90. Reflecting the work of O'Donovan, O. *Begotten or made?* Oxford University Press. 1984
[30] Wyatt, J. (2004). *Matters of Life and Death*. Leicester, England: Inter-Varsity Press. Pg 92-93
[31] Berry, C. (1999). *Artificial Reproduction*. London: CMF. P4.
[32] Monash IVF. (2011). *Donor Sperm Program: Monash IVF*. Retrieved October 16th, 2011, from Monash IVF:Life starts here: http://www.monashivf.edu.au/Services/Donor_Programs1/Donor_Sperm_Program.aspx
[33] Retrieved September 10th, 2011, from New York Times: http://www.nytimes.com/2011/09/06/health/06donor.html?_r=1

of the donor to remain anonymous versus the rights of the child to know their genetic background including potential inheritable diseases. Furthermore, having a large number of half brothers and sisters increases the risks of increasing recessively inherited diseases within a population.

Unlike gamete donation, surrogacy does not require the introduction of genetic material from another outside the marriage relationship. Rather the uterus of another woman is utilised to nurture a developing embryo (either where the genetic mother does not have a uterus, is medically unable to carry a feotus to term, or in some cases, by choice). Once delivered, the baby is 'handed back' to the 'commissioning mother'. It creates a complex relationship between the three 'parents' and the child. The physical and emotional wellbeing of the surrogate mother must be maintained. Yet weighing up the rights of the surrogate mother, genetic mother and foetus can become a very murky area. Futher exploration of the ethics of surrogacy is beyond the scope of this article.

Part 4

Pastoral considerations

There is no denying the social and emotional impact of infertility. Couples and individuals may experience a wide range or emotions including grief, jealousy, anger, depression, isolation and frustration. Unfortunately, people usually suffer in isolation from their usual means of support. IVF maintains a sense of hope through providing a potential pathway towards pregnancy. However, it comes with the added stress of financial strain, a barrage of unpleasant medical procedures and a prolonged period of pressure and stress on the couple involved.

Trusting God

Many Christians may ask what 'trusting God' looks like under circumstances of infertility. At some point couples may be required to comes to terms with the idea that biological children may not be a part of God's sovereign plan for them. Though "in all things God works for the good of those who love him, who have been called according to his purpose" (Romans 8:28), this does not mean that He will fulfil all our desires. Contentment in God alone must be sought (just as it should for couples blessed with children). Just as Paul accepted that God would not remove the 'thorn from his flesh' we should be encouraged that God's grace is sufficient for us and His power is made perfect in weakness (2 Corinthians 12:8-9). For some, the inability to have their own biological children may be an opportunity to show blessing towards other vulnerable members of our community through means such as adoption, foster care or caring for the sick and lonely. For others it enables enhanced relationships with children of other significant people in their lives, such as neices, nephews, godchildren or close family friends. Furthermore it may enable time to build other relationships that results in spiritual children and the growth of God's kingdom.

Considering adoption/foster care or other alternative nurturing roles

As Christians we should have a high appreciation of adoption; God's grace is lavished upon us that we are adopted *as sons* through Jesus Christ awaiting our guaranteed inheritance (Ephesians 1:5-14).

However we must remember that as rewarding as adoption may be, it may not lessen the grief that infertility can bring. To name but a few griefs, individuals and couples must grieve the loss of potential to have a genetically related child and to participate in one element of God's creative process. The loss of joy of having a new creation grow and form within one's womb cannot be denied. Individuals and couples may grieve the loss of social idenity or participation that parenthood brings and the dream of what their familiy unit may look like and how it may interact with the rest of their communities. Additionally they must deal with the loss of control over their bodies and reproductive capacity. This is often entangled with a person's perception of his/her identity. Adoption is also a costly and slow process with limited babies available within Australia[34]. The comment "you can always adopt" is thus an inconsiderate one despite it being a possible avenue of expression for the couple's nurturing desires.

Dealing with grief

As with all grief, the resulting emotions and variable responses may appear at different times for different people. This can bring additional strain in an already strained relationship as couples may struggle to express their feelings or may expect their partner to be able to relate and understand. Additional feelings of isolation may ensue. Additionally some may struggle with the perception that they should maintain a certain role such as the husband being an 'emotional rock' for his wife despite his own inner turmoil. The Bible passage 1 Samuel 1-2:11 provides an insightful look at the relational differences between men and women. Hannah is a faithful, God-fearing woman who refuses to worship the fertility God Baal and is tormented by her husband's other (fertile) wife. Elkanah, her God fearing husband, loves her deeply despite her being barren. Distressed by her weeping, he attempts to comfort her: "Hannah, why do you weep? And why do you not eat? And why is your heart sad? Am I not more to you than ten sons?" (1 Samuel 1:8). Though well meaning, it is almost humorous how far he misses the mark! Yet for Hannah, no doubt his comments must have stung deeply and further increased her feelings of isolation.

For some couples, pursuing IVF may delay the grieving process: there is always the hope that the next cycle will produce their desired child. There is no clear end point in the process. Couples must therefore be clear in advance what their personal limitations are (including emotional, financial, social and ethical), what things they are willing to compromise (including other relationships, work and finances) and where they will draw the line. Again these things may differ for each partner and thus clear communication and review of these guidelines will be required.

Children as idols

For both couples blessed with biological children and those without we must be aware of the real danger for children to become idols, an object of the heart that we elevate to being so central to our lives that we displace God from His rightful position. Keller defines these 'counterfeit gods' as:

> *Anything so central and essential to your life that, should you lose it, your life would feel hardly worth living. An idol has such a controlling position in your heart that you can spend most of your passion and energy, your emotional and financial resources, on it without a second thought.*[35]

---

[34] For futher information see www.adoptionaustralia.com.au It's important to note that restrictions and regulations vary from state to state.
[35] Keller, T. (2009). Counterfeit Gods. London: Hodder and Stoughton. P xviii

For a couple struggling with infertility this may be overtly seen where children are sought at a cost to the couple's relationship with God, each other and close friends. For couples with children it may be subtler to detect. The anxiety and educational investment surrounding a child's achievements and success may, for example, reflect a parent's attainment of significance through his/her child's achievements. Seeking a sense of completeness from children can result in significant pressure on children that is ultimately to their detriment. This is not to say that couples struggling with infertility should not grieve or seek medical assistance, but rather to highlight the dangers of children as idols if we (mis)place our significance in them – a danger that exists for families with and without children. Our identity should not come from our (ability to produce) children but rather from being a child – and thus heir – of God.

Financial stewards

Christians are stewards of God's money and thus have an obligation to wisely consider how they should best use His money[36]. IVF can cost thousands of dollars (even, in Australia, after a Medicare rebate) per cycle without any guarantee of conception[37]. Costs of adoption are also prohibitive for some couples. Arguably there is a difference between spending such amounts of money to bring a new child into the world and spending comparative amounts to provide care and love for an existing child requiring a family home.

How should the Church respond?

Our church communities must respond with openness and sensitivity, embracing couples struggling with infertility and affirming their valuable contribution to our communities. The couple's grief should be the church's grief. We must weep when they weep (Romans 12:15) and be sensitive to the fact that most of the time solemn Biblical truths or pious words are be better replaced with a listening ear and a tender heart[38].

We must be careful not to make the same mistake as Job's friends and offer (even well founded) advice or Biblical passages. "Just trust God" or "All in His good time" often do more to make the speaker feel better than they do for the receiver. Sometimes questions of 'why?' are better left to God's infinite wisdom. John Wyatt puts it more succinctly:

> *When confronted with human suffering, like Job's friends we frequently have an overwhelming desire to provide neat explanations. "This happened because of that… God is teaching you to…" Instead we should learn from the book of Job.*

---

[36] I do not wish to make judgements about how couples spend their money here but rather wish to point out that financial stewardship needs to be another consideration.
[37] Costs vary depending on a patient's individual circumstances and clinic attended thus I am unable to provide specifics costs here.
[38] Carson, D. (1990). How Long, O Lord. Michigan: Baker Book House. Pg 247-252

> *There can be no human explanations for the mystery of suffering – only the presence of a loving and suffering God."*[39]

Well-meaning advice is often not limited to Biblical theology but extends to (usually unfounded) medical advice. Commonly heard are phrases like "just relax" which tends only to enhance a person's stress. Furthermore, while stress and anxiety can impact a woman's hormones and natural cycle, 80-85% of women will have a medically diagnosable cause for their infertility[40]. Others may have a slight chromosomal abnormality or condition that is yet to be detected with standard medical testing. Thus not only is such advice unhelpful, in all likelihood it is also incorrect. Similarly, anecdotes about others who have overcome infertility are generally unhelpful and a source of frustration. Unfortunately there is no guarantee that what has worked for one person or couple will work for another, even if it appears the individual or couple have the same underlying medical condition. A listening ear is often more beneficial and comforting than our efforts to find the 'right' words[41].

We should also be sensitive to the language we use in church, both up front from the pulpit and within our own conversations. Special occasions like Mothers' day and Fathers' day can be particularly hard on grieving individuals or couples, and our choice of words can exacerbate their pain. Assumptions about a couple's reproductive choices should never be made. A couple without children may not, for example, be placing career ahead of family (as is frequently assumed) but rather be struggling with infertility. Questions about when a couple are going to have children, though often well meaning and asked in an effort to relate with others on a deeper level, may be a source of further hurt.

Our words and questions are often better directed to prayer to God. In the midst of suffering many of us find articulation of prayer difficult. We can take comfort that in our weakness the Holy Spirit intercedes for us (Romans 8:26). Yet what a blessing it is to also have others intercede on our behalf!

We must keep in mind that many areas of assisted reproduction come down to questions of conscience and that some may feel more freedom to move forward with reproductive technologies than others.

The difficult balance is that we also have a responsibility to nurture and shape one another in order that we can be fruitful members of Christ's body. We need to find ways of lovingly reminding people of God's unfailing love, mercy, and his multiple (though different) blessings. It may be appropriate to encourage areas of service that exist outside parenthood. This should in no ways deny the suffering and pain that lingers for infertile couples but rather affirm the valuable role they – God's beloved children – have in the growth of His kingdom.

---

[39] Wyatt, J. (2004). *Matters of Life and Death.* Leicester, England: Inter-Varsity Press. P 67
[40] Glahn, S L, Cutrer. W R, The infertility Companion; Hope and Help for couples facing Infertility, Zondervan. P 27
[41] Matthias Media. (2010, May). The eBriefing. *Making Babies: Infertility and the Ethics of IVF* (380). P 17.

Part 5

Study questions

Introduction

What might be some of the attitudes to infertility and Assisted Reproductive Technology within your church? How are these attitudes and beliefs communicated? In what ways are these helpful/unhelpful for people struggling with infertility?

Infertility in Scripture

Read:

Luke 1:5-6, 24-25

Genesis 16:1-6, 18:9-15

Genesis 30:1-23

1 Samuel 1:1-20

Leviticus 20:21

Isaiah 54:1

John 9:1-3

What can we learn about infertility in the Bible from these passages? What's God's attitude to it?

Suffering and sovereignty

Infertility causes immense suffering. How can God's love and sovereignty be reconciled with such suffering?

Read:

1 John 1:5

Deuteronomoy 32:3-4

Romans 5:1-8

Hebrews 4:15

Job 38:1-4

What might be some of the ways you may support and encourage people in your church or community who are struggling with infertility?

Respect for human life

Read:

Genesis 1:26-27; 5:1, 9:6

1 Corinthians 11:7

James 3:9-10

Ephesians 4:24

What does it mean to be created 'in the Image of God'?

Read:

Jeremiah 1:5

What does it mean to be known by God before we are born?

What are the implications (if any) for human embryos?

Life, fulfilment, and adoption

What is God's intended purpose for human life? How should this affect our attitude to infertility? What are the implications (if any) for human embryos?

Reflect on Ephesians 1:5-14. How, if at all, should our adoption as God's children impact our attitude to adoption?

Pray

Pray for those you know that are struggling with infertility. Examples may include praying for healing, contentment, trust in God and a healthy marital relationship.

Thank God that we are made in His image and he knows us.

Additional questions and points of reflection for couples struggling with infertility

Reflect on what the hardest aspect of infertility is for you.

How has infertility impacted your relationship with your partner? With others? With God?

Spend some time talking to God about any hurts, frustrations, grief, anguish or bitterness you may have felt. What things in your life give you joy at the moment? Spend some time praising God for those joys.

How, if at all, has this document challenged your views on infertility and ART? Does your partner agree? Your church?

Are you tempted to prioritise having a child above your relationship with God? Your partner?

Read 2 Corinthians 12:9. Do you believe that God's grace is sufficient for you? How are you going with contentment in the context of infertility? What would aid your contentment?

What does trusting God look like under such circumstances?

How have you grown through your struggle with infertility?

What steps have you taken to overcome infertility? What are the ethical and practical considerations that are preventing you from proceeding further? Does your partner agree with these limitations?

Can you identify ways in which your infertility has enabled you to encourage or bless others?

Glossary

ART	Assisted Reproductive Technology: Any procedure involving the manipulation of sperm, eggs or embryos to enable conception. Includes IUI, GIFT, ZIFT and IVF
Blastocyst	Early embryonic stage (reached at approximately day 5 or 6 post fertilisation).
Fertilisation	The process where sperm penetrates the egg, resulting in a zygote.
GIFT	Gamete Intra-Fallopian Transfer: Process where a retrieved egg and sperm are inserted back into a fallopian tube with the aim of achieving fertilisation.
ICSI	Intacytoplasmic Sperm Injection: The injection of a single sperm directly into an egg to achieve fertilisation. This technique may be useful where sperm count is low or sperm are immotile.
IUI	Intrauterine insemination: Specially treated sperm are inserted directly into a woman's uterus via a long catheter.
Phenotype	The observable physical characteristics of an organism resulting from the expression of his/her genes and environmental influence
ZIFT	Zygote Intrafallopian Transfer: Zygotes (produced in a petri dish) are inserted into a woman's fallopian tube where they may travel to the uterus for implantation.
Zygote	The cell produced when sperm and egg combine through fertilisation. This is the earliest developmental stage of the embryo.

Bibliography

- Berry, C. (1999). *Artificial Reproduction.* London: CMF.
- BioEdge. (2011). *BioEdge: bioethics news from around the world.* (M. Cook, Editor) Retrieved September 10th, 2011: http://www.bioedge.org/index.php/tools/newsletter_/
- Carson, D. (1990). *How Long, O Lord.* Grand Rapids, Michigan: Baker Book House.
- Chenoweth, D. J. (2011). What is it to be human?: The view from IVF practice - a personal reflection. (J. Foley, Ed.) *Luke's Journal , 16* (2).
- City Fertility Clinic. (2009). Retrieved September 15th, 2011, from IVF Process - IVF Treatment: http://www.cityfertility.com.au/IVF-Process
- Fisher, A. (1989). *IVF: The Critical Issues.* Melbourne: Collins Dove.
- Geisler, N. L. (2003). *Christian Ethics.* Grand Rapids, Michigan: Baker Book House.
- Glahn, S. L., & Cutrer, W. R. (2004). *The Infertility Companion.* Grand Rapids, Michigan: Zondervan.
- Goldingay, J. (2008). *Psalms 90-150* (Vol. 3). Grand Rapids, Michigan: Baker Academic.
- Gurevich, R. (2011). Fertility. Retrieved September 5th, 2011, from About.com: http://infertility.about.com/od/infertilitytreatments/ss/ivf_treatment.htm
- IVF Babies. (2006). *IVF Babies - Information and Resources for parents of IVF babies.* Retrieved September 6th, 2011, from http://www.ivfbabies.com/
- Keller, T. (2009). *Counterfeit Gods.* London: Hodder and Stoughton.
- Kovacs, G. T. (2003). Embryo donation at an Australian university in-vitro fertilisation clinic: issues and outcomes. Retrieved January 27, 2012, from The Medical Journal of Australia: http://www.mja.com.au/public/issues/178_03_030203/kov10329_fm.html
- Matthias Media. (2010, May). The eBriefing. *Making Babies: Infertility and the Ethics of IVF* (380).
- Meilaender. (2009). *Neither Beast Nor God.* New York: Encounter Books.
- Meilaender, G. (2005). *Bioethics.* Grand Rapids, Michigan: William B. Eerdmans Publishing Company.
- Monash IVF. (2011). *Donor Sperm Program: Monash IVF.* Retrieved October 16th, 2011, from Monash IVF:Life starts here: http://www.monashivf.edu.au/Services/Donor_Programs1/Donor_Sperm_Program.aspx
- Monash IVF. (2011). *Monash IVF Resource Library.* Retrieved October 5th, 2011, http://www.monashivf.com/InformationLibrary/Why_aren_t_we_pregnant_.aspx
- Mroz, J. (2011, September 5th). *One Sperm Donor, 150 Sons and Daughters.* Retrieved September 10th, 2011, from New York Times: http://www.nytimes.com/2011/09/06/health/06donor.html?_r=1
- Piper, J. (2007, April). *Single in Christ: a Name Better Than Sons and Daughters.* Retrieved November 2011, from Desiring God: http://www.desiringgod.org/resource-library/sermons/single-in-christ-a-name-better-than-sons-and-daughters#/watch/full
- Rachels, J. (2007). *The Elements of Moral Philosophy.* (S. Rachels, Ed.) New York: McGraw-Hill.
- Stott, J. (1994). *The Message of Romans.* Leicester, England: Inter-Varsity Press.
- Wright, C. (1996). *Deuteronomy.* (R. J. Hubbard, Ed.) Peabody, Massachusetts: Hendrickson.
- Wyatt, J. (2004). *Matters of Life and Death.* Leicester, England: Inter-Varsity Press.

ACORN PRESS LIMITED

Talking about Ethics: Negotiating the Maze
By Justin Denholm

RRP $14.95
ISBN: 978-0-9871329-1-8
72 pages
Includes individual & group study questions

What a useful book! Justin not only explains the issues, but also gives practical advice for thinking, discussion and action.

Peter Adam, Principal, Ridley Melbourne

This remarkable book is unlike any other on ethics. Reading it won't make you a better person. It won't even tell you right from wrong. But it will give you wonderfully clear and accessible wisdom how better to understand and talk with others about how we should live.

Robert Forsyth, Bishop of South Sydney, Anglican Diocese of Sydney

I warmly commend *Talking about Ethics* because it does what the title says – it talks about ethics in a way that is accessible, useful and easy to understand.

Katy Smith, Bible College SA

Justin's book brings Scripture, wisdom and experience as a doctor, ethicist and mentor into a package of helpful resources for individuals and small groups. It meets its aim: to enable Christians to converse constructively with each other and non-Christians about moral issues as a form of ethical apologetics.

Gordon Preece, Director, ETHOS

I felt challenged to think critically about my view on any number of issues, and to better articulate those views with grace and godliness to those around me.

John Quinn, Dean of Residents, New College Village, University of NSW

Justin Denholm's book is short, sensitive, sensible and scripturally engaged. For Christians concerned with ethics and with helping others think about it, here is a great place to start.

Andrew Reid, Holy Trinity Anglican Church, Doncaster VIC

If you didn't know where to start when a friend asks you about some moral dilemma, I can think of no better book to begin the process of thinking ethically so that you might better respond to the challenges of moral decision-making.

Glenn N. Davies, Bishop of North Sydney, Anglican Diocese of Sydney

---

To purchase your own copy of *Talking about Ethics,*

order through Acorn's website: www.acornpress.net.au

or

ring Acorn's office on 03 9383 1266

or

email orders@acornpress.net.au

www.ingramcontent.com/pod-product-compliance
Lightning Source LLC
Chambersburg PA
CBHW080859170526
45158CB00009B/2776